Mustapha Guenaou
Abderahim Guenaou

The notion of water in Ain El Hûts (Tlemcen)

Mustapha Guenaou
Abderahim Guenaou

The notion of water in Ain El Hûts (Tlemcen)

ScienciaScripts

Imprint

Any brand names and product names mentioned in this book are subject to trademark, brand or patent protection and are trademarks or registered trademarks of their respective holders. The use of brand names, product names, common names, trade names, product descriptions etc. even without a particular marking in this work is in no way to be construed to mean that such names may be regarded as unrestricted in respect of trademark and brand protection legislation and could thus be used by anyone.

Cover image: www.ingimage.com

This book is a translation from the original published under ISBN 978-620-6-72460-5.

Publisher:
Sciencia Scripts
is a trademark of
Dodo Books Indian Ocean Ltd. and OmniScriptum S.R.L publishing group

120 High Road, East Finchley, London, N2 9ED, United Kingdom
Str. Armeneasca 28/1, office 1, Chisinau MD-2012, Republic of Moldova, Europe
Printed at: see last page
ISBN: 978-620-8-23375-4

BY WAY OF PRESENTATION

For decades, the local history of Ain El Hûts was the subject of research by a teacher and then by his former pupil. In the same context, we would like to pay tribute to our beloved schoolteachers: the teachers who taught at the mixed school in Ain El Hûts for many years before it was renamed after a chahid, Sid Ahmed Khiat Maazouzi.According to local elders, this locality, which is socially, culturally and historically unknown to some, has over twelve centuries of history. Few geographers have visited it, with the exception of El Bekri.The collection of documents, together with a compilation of published texts, focuses mainly on a specific and singular theme: the locality of Ain El Hûts which, along with other localities including W-zidan (Ouzidan, a Frenchisation of the toponym), represents the hawz of the medina of Tlemcen, the ancient capital of the central Maghreb. These two towns, which were already populated at the time of the founding of the Abdalwadid and then Zianid kingdoms, were founded more than twelve centuries ago. Known by their respective place names, Ain El Hûts and W- zidan were visited by a wandering saint and poet. This was Sidi Ahmed Ben Youssef, a local patron saint of the town of Méliana. At the end of his visit, dating from the 15th century, this saint bequeathed a few satirical sayings, as he had done for the town of Tlemcen.

For Ain El Hûts, he is quoted as saying:

"Ain El Hûts

Drink some water and pass (^{1}) "

In our opinion, these two localities are the only ones to represent the history, culture and even the memory of the extra muros, defined by the hawz territory as opposed to the medina and the r-bad. This term was introduced into the medina, which had lost its rural character since the thirteenth century with the arrival of the first Andalusians, at the request of the first princes of Tlemcen.

In our humble opinion, these documents, both old and recent, constitute a non-exhaustive database for possible study and bibliographic use by students and teacher-researchers. They undoubtedly contain sociological, anthropological, historical and even ethnographic information.

Containing real details, these articles remain sources to be exploited to understand the history of Ain El Hûts, "blèd Eshorfa w -l-m-rab-tine" (the land of nobles and saints!).

Examination of the content of these documents leads us to a reading of the introductory sources for any research into semi-rural sociology in relation to urban sociology of the medina type: Tlemcen, an Arab-Moorish town, an Arab-Muslim town and a pre-colonial town. The distinction is made on the basis of the foundation of the hawma with its traditional road network: the derb, the driba, the s-wiqa, and so on.

The city of Tlemcen is characterised by the coexistence of two hawmats (traditional districts): the hawmat of the hadars and the hawma of the kûlûghlis. The latter dates back to the arrival and settlement of the first Ottomans, who married Arab girls whose descendants were known as the Kûlûghlis. This ethnic term refers to the children of mixed marriages between Ottomans and local women.

Within this purely simple and unambitious framework, we are sharing these information documents which we offer to users such as teacher-researchers, students, independent researchers and laymen interested in local history. These sources are written in French. We have limited ourselves to French references on the subject in order to be able to talk about writings, some of whose authors are Europeans who lived in Tlemcen for a long time. Perhaps these documentary sources will later be translated or published. In In the same vein, we intend to popularise other documents relating to Ain El Hûts, such as hawfi texts (anonymous women's poetry). Our intention has never been an ambitious compilation, but rather an opportunity for readers, enthusiasts and laymen alike, to learn more about the history of this locality which, in addition to the third day of Eid el Fitr, Eid el Ad-ha and tsashwisha, used to bring together women to go out in the fields during the spring (ziara and others) and boys for the summer (bathing at Tsahamamits). We were very saddened by the disappearance of the epitaph of Sidi Slimane, who was the brother of the founder of the city of Fez (Morocco). Then, we boast about the quality of the work done by our masters, our elders and our close friends.

WATER (EL MA/ EL MA-E)[2]

Water is a necessary and indispensable liquid for the life of human beings, animals and vegetation; but for its quality, it has several symbolic and spiritual meanings. These meanings are grouped under three and several important and dominant themes:

1. Water is a source of life,
2. A food centre,
3. A regenerating centre,
4. Water, a means of purification and cleanliness,
5. A danger for the careless. This is the case with bad weather and natural disasters (heavy rain and deluge).

It represents infinity and clarity, promising development to avoid any threat of resorption, impotence or even disappearance. In this case, the spring is associated with bathing and Moorish baths, or even ablutions. The spring, in general, brings water, life, strength, hygiene and cleanliness to social life, as well as to the body and spirit, and, as a form of animation and manifestation, is the instrument of purification. body and ritual. Ablutions play an essential role in this.

For Muslims, the holy water of 'zemzem' is a symbol of the sacred for making a vow, since purification is achieved through clean water.

In the ablutions, the faithful consecrate and sacrifice themselves for the ritual, which is both purifying and regenerative.

Its nature gives it the quality of drink and also leads it to purification. Water takes on the role of a source of life to regenerate a bodily and spiritual situation. To remain the symbol of purification, fertility, purity and divine grace, its virtue introduces wisdom and benefit. Its fluidity signifies its abundance, since it represents the possibilities of great and small manifestations. Water is the origin of life, and conveys it in a vital atmosphere.

If water is a source of life, it can also be the source of death. Moreover, water, as a constructive and destructive creator, is used in the first days of birth and is recommended for washing the body.

Water is associated with the function and virtue of the well, the rain, the river, the spring and the fountain, the places where this drink is offered, appreciated by the Koran as well as by prophetic tradition. Water is also a source of joy and cleanliness, even a sport and a source of wonder. Sacred places are sometimes found very close to the water and play an important role in therapy.

Without water, life is no longer a source of contact and growth for human beings, animals and plants alike.

In demand, water is a lifesaver in remote, deserted places.

Alteration water is a means of food and survival.

In the Muslim religion, it is requested for prayer and becomes a necessity and a supplication. A fertiliser for the soil, it expresses the Almighty's benevolence and joie de vivre.

In Muslim traditions, water is presented to all visitors and guests to ensure hospitality and peace at rest.

It takes the form of a sign of blessing from God, the Creator of the universe, to symbolise the Power of God.

Water is the origin of divine existence in Muslim belief. More importantly in Muslim belief, water represents wisdom and spiritual life.

The water of life is living water and the source of growth.
It will purify the man, woman, man or child, just as it can heal him for the rest of his life. Then, she is called upon for an introductory washing into eternal rest. It represents lucidity, transparency, fecundity and preference in choice, since it possesses and represents the virtue of purification and sacralisation, confirming its use for ablutions before any prayer.

In a state of defilement, its negligence undoubtedly entails a religious offence, not accepted by Islam.

By its destructive function, it can erase all traces of human, animal and plant presence, a s well as witnesses to History, Memory and Memories.

In the Holy Places of Islam, the presence of water is the object of pilgrimage because the source of water is a psychological remedy and a ritual in religion.

Just as water has a therapeutic power, it can also have a destructive and punitive power. Water can be fresh or salty.The first symbolises purity and cleanliness, while the second is salty to represent seawater, clean for hygiene and purification but difficult to drink.

As for bitter water, it produces a curse that sometimes ends in the death of the person who consumed it. Rough water is a sign of misfortune, representing evil and disorder. As for calm water, it signifies peace and enjoyment in a situation of discipline and order. Rain, as water, which also has the properties of water, is blessed because it comes from the sky, one of the divine signs recognised by Islam.

It is found in the streams of Paradise, near which the houris live. These streams are special in that their water is blessed and alive; water is the Divine Essence, since it fulfils the functions of creation and growth, symbolising hygiene, cleanliness and purity, and even the purification of body and soul for all prayer in the Muslim religion.

Prayer is only validly performed and accepted on the basis of bodily purification and purification of the soul.

Water symbolises life on land and in the sea. Rainwater is celestial for the fertility of the land and the fecundity of the living. Fertilisation is triggered under the right conditions, as in the case of germination within the earth. Water fertilises the earth and becomes a vital and growth force, becoming water, the sign and symbol of energy.

Cleanliness attracts people, but unfit or dirty water remains a symbol of stench, filth and horror, and pollution has only brought disaster, disease and death.
In the sacred book, water fulfils a number of functions, linked to purification, growth and development, as well as being a tool for disaster and catastrophe.

As a means of purification, defilement and impurity will only disappear through the use of pure, clean water.

While ablutions are required, purification remains the key to all prayer. To do this, you need to purify your body and mind. Islam has always attached great importance to water, seeing it as a factor in fertility and growth.

Water is a source of life and a return to life, especially for plant reproduction. In the Hereafter, it will become an instrument of torment as it will be boiling, tearing and foul. Good water is a symbol of joy, of a happy life, because of its clear, pleasant and delicious taste. It therefore represents the benefits that God bestows on Muslim believers.It symbolises the manifestation of God, and will remain fertile water for the germination of seeds and flowers, for the purification of the body, the place and the soul, and as a lustre to bear the sign and emblem of all living things.

Source : El Ain El Kbira

The small village of Ain El Hout, around eight kilometres north of Tlemcen, is made up of two groups of inhabitants: one situated to the east, around the tombs of saints and a spring which serves as the village's eponymous spring; the other called Tghalimet, situated a little further west, is populated by farm workers, and does not seem to benefit from the blessing of the marabouts. The village notables, descendants of the marabouts, claim descent from Soleiman ben Abdallah, brother of Idriss, the founder of Fez, the first sultan to seize the territory of Tlemcen, then called Agadir (in the 10th century). Oral traditions are eminently respectable, which is why we will not contradict them. But the truth compels us to say that the written texts, which are our only archives concerning the village and which can be found near the tombs of the marabouts, date back to the 18th century.

When you get to Ain El Hout on the road that leaves Bab el Kermadin ("the door of the tile-makers"), which crosses the stream called Chabet el Horra on the road from Bréa to Négrier, and which brings you to the village about two kilometres from this junction, the first thing that catches your eye is a very well-maintained pool, much frequented by Muslim women and girls, populated by a multitude of goldfish.

The spring that feeds this pond gave its name to the village. A sign informs passing tourists that it is forbidden to touch the fish. The legend goes as follows:

"One day in May, just as the weather was inviting, a sister of the Choumissa princess set off to pick flowers in the outskirts of Tlemcen and wandered along the banks of the Chabet el Horra.

A handsome young man spotted her, noticed her beauty and wanted to see her more closely. He complimented the princess and offered to help her pick some flowers. The princess suspected impure intentions in the young man and ran to put a respectable distance between herself and him. The race lasted a good half an hour and the princess was dreading the moment when she would be caught by the young man, when she saw a large spring ahead of her. She had no time to think and only made a vow to escape her pursuer. Immediately, she threw herself into the spring and God, who always sympathises with the miseries of the weak, immediately changed her into a fish". It is obviously out of respect for this beautiful princess that it is still forbidden to fish in the village pond.

THE PUBLIC FOUNTAINS OF AIN EL HUTS[3]

The water main from El Ain El Kbira to the Dar El Arça district. Four public fountains will be built, in this order:

Saqayèts El Djamaa

This fountain is called Saqayèts El Djamaa by some and Saqayèts Hammam Es sadaq by others[4] . It was located at an equal distance between the Moorish bath of Sadaq Benhamou and the old mosque of Ain El Houts. When his house was built, ould Berrezoug moved it to the south at a distance of around five metres. Today, it stands at the foot of a monument to the martyrs, on the site of the former barber-shop of Ghalem. For the record, this fountain was used to fill the communal pool used to irrigate the orchards in the outbuilding. In my childhood, this pool was used as a children's swimming pool. This same pool was located between two former cafés:

- Qahwat El Hammam, with tables and chairs
- Qawat El Bilem, with the mats of the Beni Snous, commonly known as h-sayèr.

Saqayèts Sidi Mansour

Because of its proximity to the mausoleum of the saint Sidi Mansour, this fountain has a number of distinctive features that set it apart from the usual fountains such as Saqayèts El Djamaa or Saqayèts Hammam Es sadaq. It is a washing and drinking fountain with a triple function:

- Drinking water supply for women in the adjoining and upper blocks[5]
- The women's washhouse on the same islets
- The drinking trough of animals[6] and herds[7]
- Irrigation of orchards of dependency.

For family and domestic consumption, children, women and girls volunteered to supply the family with drinking water. The local population used the popular expression "yamlaw el ma" for some and "y'aamro el ma" for others. For us, these are water carriers for family consumption. For buckets, people speak of "byadène" (plural of bidoun) instead of delw.

Many women, who worked with wool during the war of national liberation, came to wash quantities of wool to be used for drying, sprinkling with kebrits (yellow sulphur of the Qannout type for preservation and protection against insects) and other treatments. mites), tsqardish (wool carding), ghzil (spinning) in order to transport the finished product for sale at Sûq El Ghzel (on the square known as Mawqaf) in Tlemcen. Among the women I had seen were my paternal grandmother and my aunts, who carried rezma essôf (woollen bundles) on their respective heads as far as Tlemcen, and especially when walking to their destination.

A few years after the outbreak of the War of National Liberation, it became very difficult for women who produced woollen products to get around Tlemcen on foot.

Ould Melouka would come to collect wool. Instead of the collection hired by the buyer, the women came to offer him their respective products:
Dar Boumedienne Benguedih, Hadj Mansour's cousin, opposite Ahmed Laddawi's communal oven, owned by Ahmed Merzouqi, whose real name was Chergui and who was the father of Ammi Abderahmane Ettarrah.

Other women came to the washhouse to wash clothes, various blankets, other bedding and bedding, etc. At the end of the fifties (20th century), we noted the markers of the :

- Contribution

- Solidarity

- Physical participation

- Use of assistance

- Mutual assistance

- Competitions

- Support.

This heptaptych takes us back to the markers of what is known in the language practices of the locality as the 'Touiza'. The work was carried out in good conditions, animated by a feminine and particular atmosphere, since we heard the women sing an anonymous feminine poetry, commonly called 'Touiza'.
"Hawfi. The farmers, who included owners of land, fields and orchards, as well as those who hired out their labour, got up early to go to their place of work. Owners of cattle and herds would stop off at the trough while the animals and herds drank in peace and quiet.

Passing in front of this fountain, these peasants drank this fresh, clear water from the Ain El Kebira early in the morning or filled their respective pottery jugs (q- lèl, plural of qolla), covered with jute (khish or shkara) interspersed w i t h wheat grains to keep the container and contents cool during the summer.

Saqayèts Benqrich

This simple public fountain, with its customary bezzûl essaqaya (spigot), formed a permanent flow tap. It supplied the houses on the block where the Benmansour families and their El Abbes cousins lived, including Si Abdelazziz Benmansour, a landowner and former town councillor. His house is opposite, two or three metres away.
For the record, Ammi Abdeldjelil El Abbes, landowner and bearer of a baraka, a favour towards the socio-cultural practice commonly known as "tsa'zima".
We called him "Ammi Abdeldjelil El 'Azzam", as did Ammi Belkhouane, whose real name was Khouane Bounouar, and Ammi Gryd, whose real name was Baraka. The latter two are followers of the Aissawiya tariqa.

All three took water from a glass and recited a few Koranic verses in the same container for a token sum, in return for reading the sacred verses. For some it was holy water, for others it was sacred water, because the water was purified by the Koran. The sick, affected by El lezm (tonsil ganglions) or other facial ailments. This water is commonly known locally as "El ma Mraqi":
The belief of the local population refers to markers of mental acceptance and moral satisfaction in the context of the healing of evil by the Koranic verses. The people of Ain El Hûts believe in this socio-cultural practice for the following reasons and motives:
- Healing through Koranic verses or the sacred Koran.
- Healing favours granted to specific individuals.
- The mentality linked to popular belief.
- Acceptance of the idea of purifying water to become similar to Zemzem's holy water.

By way of illustration, let's take the example of Ammi Abdeldjelil, known for this socio-cultural practice and the favour of the baraka that God is said to have given him, for a very long time. When children went to see him for what we call 'El ma Mraqi'.

According to local tradition, Ammi Abdeldjelil had the "Azima' for some and 'hikma' for others. Moreover, he would half-fill the glass, to avoid any spillage of a few drops of water blessed by the Koranic verses. It was advisable to take care of the contents of the glass of water along the way. He would fill the glass from the fountain known as Saqayèts Benqriche.

Water flows abundantly and uninterruptedly, as a small stream used to lead to the destination of the verdant orchards and fruit trees, including orange and olive groves. The water from Saqayèts Benqrich is used to irrigate several orchards, renowned for the diversity of their fruit.

For further information, Saqayèts Benqrich was adjoining and behind the house known as Dar Esharef, a great patron of the arts and father of our classmate at the Lycée Docteur Benzerdjeb, Brahim, who became an executive in an administration. He is now retired.

A large room was forcibly occupied by the SAS (specialised administrative section) administration. This room was used as a headquarters and place of torture for the local male population, affiliated to the FLN or ALN, until the construction, between 1956 and 1958, of the military fort, commonly known as "El Bordj", an enclosure that covered the three sides of the mausoleum of Sidi Abdellah Benmansour, the eponymous saint and patron saint of Ain El Hûts.

Saqayèts Dar Esheikh

This fountain is a drinking fountain because the peasants of Dar El 'Arça, a neighbourhood of descendants of local saints such as :
- Sidi Abdellah Benmansour.
- Sidi Mohamed Benali.
- Sidi Slimane.

As with other public fountains, this fountain has three traditional functions:
▪ Drinking water supply for the women of dar El 'Arça
▪ The watering place for cattle and herds, belonging to the families known as El m'rabtine
▪ Irrigating the orchards in the outbuilding below the Dar Esheikh house
▪ Supplying water to the Moorish bath, formerly known as Hammam Dar El Arça, a former Moorish bath built in Ottoman times.
For the History and Memory of this public fountain or drinking fountain is currently located two metres from a historical wall reminding us of the spectacle

of the religious brotherhood of the Aissawa for some and tariqa Aissawiya for others. Every third day of Eid, whether it's Aïd Seghir (Aïd El Fitr) or Aïd El Kbir (Aïd El Adha), the followers of the Aissawa, who come from all over western Algeria, gather to celebrate the festival. Every year, followers of the religious order would form a group of more than two rangers for the procession from Ain El Kebira. These followers, popular and unusual dancers, were always accompanied by musicians or players of instruments such as :

- Tbèl , a drum
- Ghaïta for some and zorna for others
- Bendir for some and bendayer for others
- Tbila or neqarèts.

The procession starts from the Ain El Kebira (main spring) to the place known as Qawç for some and majmaa Eçollah for others. After a stop and a show in front of the mausoleum of Sidi Mansour, the troupe breaks up to climb to the mausoleum of Sidi Abdellah for a ziara, a pious visit to the burial place of the eponymous saint and patron saint of Ain El Hûts. It's a high-speed race: the first to arrive is the follower with the most merit and baraka.

After the pious visit, the devotees, followed by children and the male population, took the same route to the mausoleum of Sidi Mohamed Benali for another pious visit. The families attended a ceremony known as "El Heurrif", an auction of products collected along the procession, offerings of bread, dates, fruit and other consumable and, above all, saleable products.

At the end of the ceremony, the proceeds would be given to the mqaddem to supplement the money collected along the procession. The event was painted by a European painter, André Suréda (1872-1930), who attended the procession between 1910 and 1920. After the two pious and respective visits to the two burial places of the saints, the followers of the Aissawa and the local population and guests descend a path, described as a rough descent, to the esplanade of Dar Esheikh, near the public drinking fountain.

Under the large mulberry tree, in the shade, the devotees settle down for the big show, justifying the end of the ceremony and the procession. Those present are asked to stand on a wall so that they can better observe the public spectacle, with great respect and exemplary discipline. The end of the show announces the rendezvous for the next Eid festival.

Saquiet Ethahtania (tsaa Dar El Arça)

To get closer to the water source, the families living a few metres from the place known as Khrareb, the former site of the horsehair factory, commonly known as Rhat Benammar, which owned the neighbouring land, decided to extend the water pipe from the Dar Esheikh fountain and drinking trough. They chose an entrenched spot to install a fountain. According to an inscription (14.4.64) on the wall adjoining the fountain, it dates from 14 April 1964. This date mainly refers to the beginning of national independence. Designated as such, this fountain also served as a drinking fountain for the local farmers, who would head out early in the morning to their fields and orchards in Tsghalimet and Sennoun.

10 August 1977, photo taken near Ain Bent Essoltane by Dr Boumediene Bouali

THE SPRINGS AROUND AIN EL HUTS

1- Ain Seb'ûn

It is located not far from Bene eddib and Ennekhla.

2- Ain Bent Essoltane

It is located on the banks of El Mechraâ de l'Oued, near the old water mills.

Rainwater in the Koran

The people of Ain El Hûts recommend using rainwater collected in containers. In fact, several verses in the Koran refer to the virtues of water.

The use of water as part of Tahsin El Beyt

The people of Ain El Hûts recommend this water to protect their homes from occult practices and the evil eye.

Water from source

Before drinking spring water, it is advisable to remember El Besmala

"Bismi Allah

The container

In Arabic, several words are used to designate a container:

Arabs are used to storing and/or transporting water in an animal skin sack (goat skin, etc.), commonly known as a "distinguée".

- "guerba" for water
- "Shekwa" for milk and whey
- "Mezwed" for flour, ground wheat, ground barley
- "Eukka" for honey

- "Zokra" for vinegar

Source :Lissan El Arabe

Forgotten Arabic names for rain in Ain El Hûts

Rainwater was known by a multitude of names. The vocabulary that has fallen into disuse is :

"El haya-e" for rainwater, having revived the dry land

"El Ghayth" for the rain that fell after the drought

"Eddima" for the rain that lasts

"Eshabibou" for the rain that falls repeatedly

"Errihma" for the lightly falling rain

"El Quitquitou" for rain drops

"El Wabel" for the big drops of rain

"El 'Ayn" for non-stop rain for days on end.

"El Walyû" for a rainy season

"ES Sayiatsatsôu" for a downpour that washes away whatever it finds in its path

Source : Fiq-h Ellougha/ Ethaaalibi

Names for rain in Arabic

In Arabic, rain has several names, including :

"El Houtsane" for moderate rainfall

"El Ghoudq" for a shower

"El Ya'loul" the rain that falls and stops to fall again

"Esh Shou'eboub" for the great downpour

Other rain-related vocabulary

In Arabic, several terms are used to describe rain. By way of illustration, we could mention :

"Ettôl" to designate light rain

"El Wadq" to designate non-stop rain

Snow vocabulary

In Arabic, several terms are used to describe snow:

"Essiqtou

"El Ghourabou

"El Khashifou

"Dahko

"El Djamdou

"El Khashafou

"El Houlhoulou

"El Damqou

"Eddarribou

"El 'Adrassou

"Eddamikou

"Eddalmou

Source :Lissan El Arabe

El Oued: a special designation

Surrounded by orchards (fruit trees), the Ain El Hûts region is dedicated to agriculture. The local population refers to El Oued as the great river, which originates at El Mefrouche, passing through Ain Fezza, Saf Saf, before flowing into the Tafna at Sekkak.

El Oued collects water from :

- Low-flow springs along the river
- High-flow springs along the river
- Gueltas (Tsamlallou, Tbal, etc.)
- Ain Gueroui (not far from Mohamed Bekhti's orchard)
- And so on.

Tradition of boiling water, a practice of yesteryear

The people of Ain El Hûts used to use a kettle, commonly known as a "Kafatsira". As far as etymology is concerned, this term has its origins in the French word cafetière. This is the case of the language practice of the Algerian speaker.

Out of the water: names and origins

Rich in vocabulary, the Arabic language uses words, and more particularly verbs, according to their source:

"Sahha", water from the clouds

"Naba'a", water from a spring

"Inbajassa", water from the rocks

"Fada", overflowing water, from the river/ a river

"Wakaffa", water (drops, etc.) from the ceiling

"Saraba", water from an animal skin pouch (from a crack and so on)

"Rashaha", water from a container (from a crack and so on)

"Isakaba", water from a spring

Source : Fiq-h Ellougha/ Ethaaalibi

The progression of thirst: growth and successive stages of thirst

In Arabic, the dictionaries remind us of the degrees of thirst that we represent graphically by a pyramid

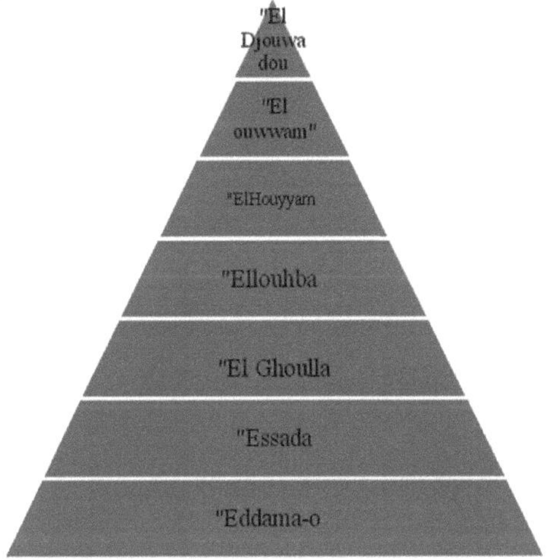

Source : Fiq-h Ellougha/ Ethaaalibi

Hfaynats El Ma

In the language used in Ain ElHûts, the local population uses a vocabulary to refer to the quantity/capacity of water:

- "Djoghma" for a sip of water
- "Hfina" to designate a quantity of water contained in the palm of the hand (the inside of a single hand).
- "Hafna" to designate a quantity of water contained in the two palms of the hand (the inside of the two hands together).

The names of l'eau

Several words are used in Arabic to describe water:

"El Abab

"El Hassir

"El Qasmou

"El Kawkabou

"El Balal

"Er Riq

"Er Red-'a

"Etsa-emour

"Er Rida'mane

Source "Mou'djam esma'e El Ashya-e

Consumption of hammam water and containers

In the ancestral tradition, the local population of Ain El Hûts uses water containers:

- "tassats El Hammam
- "Stela", small container for ablutions
- "El Bidoun" (origin: French Algerian): the barrel
- "Qbiba: a container with handles for carrying water
- "El Eub": a water container for bathing in the hammam.
- "El Borma" to designate the water collection basin
- "Brimats El Hammam", the container, in the form of a can, for transporting small quantities of water.
at the bain maure

Guelats El Oued

In Ain El Hûts, the local population uses the word
"Guelta" to designate :

- An open-air water bowl
- A depression where a quantity of water accumulates
- A pool of water

Locally, the term is used without distinguishing between the areas where the
water accumulates, although the depth varies from one place to another:

- Deep
- Shallow
- Superficial (surface)

Water consumption: use of the heufna and others

In the local tradition of individual water consumption, people use :

- "Hfina", with the use of the hand
- "Heufna", with both hands raised
- "Tsakri'âa", using the lips of the mouth.

Utensils for drinking water

In Arabic, the variety of containers depends on the material used to make them:

"El Qad-h", made of glass

"El 'ôsso", made of wood

"El M'rkane", made from earthenware

"Essowwa'e", made of silver or gold

Water from the fountain and others

At Ain El Hûts, the name given to the water depends on where it comes from:

- "El Ma Tssa El 'Ain", coming from the spring
- "El Ma Tsaa Es Seqaya", from the public fountain
- "El Ma Tsaa El Bir", from the well.
- "El Ma Tsaa El Oued", from the Oued (large river)
- "El Ma Tsaa Saqiya", coming from the stream
- "El Ma Tsaa Ennow", rainwater
- "El Ma Tsaa Entaf", waste water

The wells

In Arabic, wells are differentiated by their respective names

-" El Qalib", a well missing its owner

-" El Djob", the deep well

- "Er Rakiya, the low-flow well

- "El 'Alim", the high-flow well

- "Erras, the great well

-" El Khassif", the well dug using stone (ancient tradition and work mouen)
Source : Fiq-h Ellougha/ Ethaaalibi

The difference between types of glass at water

The Arabic language distinguishes between a glass with a ring and one without:

- "Kouz", an ance glass

-Koub", a glass without ance

The bucket o f water from .

The population of Ain El Hûts had known several forms of seal or can:

- "Bidou El Marikane", American seal, used after 1942
- "Edlou", the seal used to draw water from the well
- "El Bidoun Esghir" , small seal
- "Bido el Blastic", plastic seal
- "Bidoun El Hdid, metal seal

RECOMMENDATIONS FOR STAYING HYDRATED[8]

In Arab-Muslim culture, the people of Ain El Hûts have inherited instructions and recommendations for keeping men, women and children healthy by keeping the human body hydrated. To this end, five recommendations remind us of the need to drink water:

1- Early in the morning
2- Before any meal
3- Between meals
4- Before entering the hammam
5- Before going to bed.

It is important to note the place of the number five in local culture and traditions, and to be able to understand these recommendations, which go back a long way in the socio-cultural practices of the people of Ain El Hûts. The daily recommendation to drink water applies to all people, except adults, during the Lenten period of the holy month of Ramadhan and the days established by the Sunnah. We had to wait several years to find an explanation, confirmed by the WHO (World Health Organisation), which recommends a minimum consumption of two litres of water a day. Our curiosity leads us to differentiate between three categories of people:

- The women.
- The men.
- The children.

According to local tradition, the people with the most to gain from drinking more than two litres are in this timeline:

- Babies.
- The children.
- The sick.
- Old people.[9]

Science has provided us with knowledge about water consumption to remind us of the water requirements of the human body. It turns out that the human body contains a large quantity of water. It is a necessary liquid. This is explained by the results of scientific research.

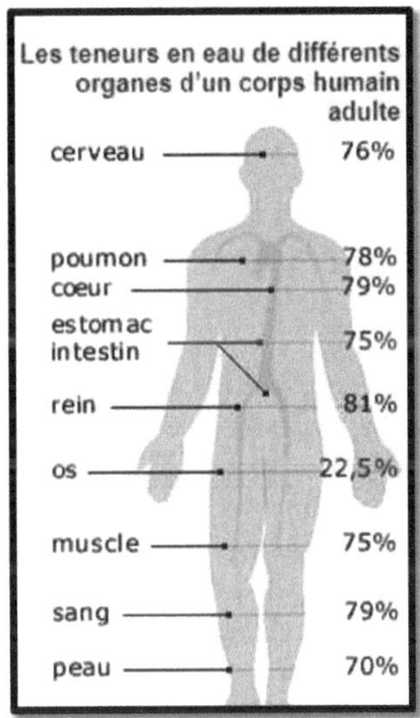

Source: cnrs.fr

The question of water is a pertinent one, which refers us mainly to the organs, which are considered to be very rich in water:

- The heart.
- The brain .
- The kidney.

The human body requires a certain amount of water to meet its needs. Moreover, these needs lead to scientific research and the results of academic work:

The human body demands water and eliminates it gradually during the day by :

- Urine.
- Sweating.
- Others .

In this context, we can speak of an exchange carried out by the human body between the acquisition and loss of water.

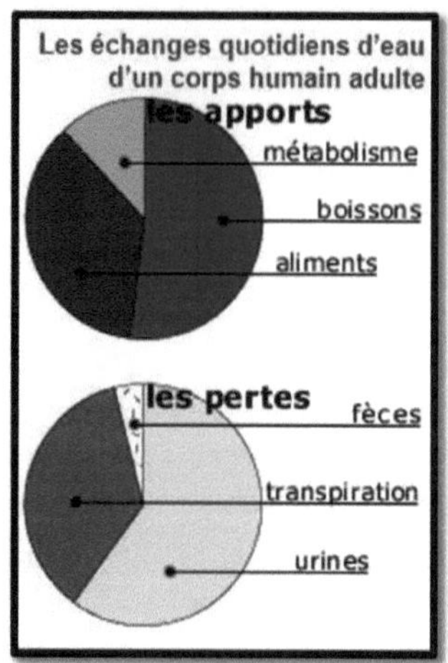

Sources: cnrs.fr

PROHIBITIONS SOCIO-CULTURAL

In practice, grandmothers and mothers forbade children to drink water standing up and recommended sitting down. The reasons for this are to be found in the Sunnah :

In Ain El Hûts, several practices are socially and culturally forbidden. These socio-societal practices are referred to by socio-anthropologists as socio-societal prohibitions.

Here are a few examples:

- Pissing on/in the water
- Striking the surface of the water with blows

The local population recommends respecting the prophetic hadiths (sayings/sayings/practices):

- Sitting down to drink
- Do not drink while standing
- Drinking water intermittently (with a momentary pause)
- Drink three times per intermittently (with momentary stop)

Drinking water with meals

We have noted a number of comments and observations relating to water consumption:

The people of Ain El Hûts recommend :

- Drink a glass of water every morning
- Do not drink water with meals
- Drink water after each meal
- Respect one third of the water capacity of the belly
- Drink the water known as "El Ma M'reqqi" or the water from Mecca "Ma-e Zem Zem".

BIBLIOGRAPHY

Sacred books

1-The Koran
(Translation new by the Sheikh Boubakeur Hamza)
Algiers, ENAG éditions, 1989 (2 volumes)

Dictionaries, Glossaries and Encyclopaedias

BARAKE(Bassam)
Linguistic dictionary. French - Arabic
With an alphabetical index of Arabic terms
Tripoli(Lebanon),Jarouss Presse, (?), 298 p

BEAUSSIER(Marcelin)
Arabic-French Dictionary New, revised, corrected and updated edition
expanded by Mohammed Bencheneb Algiers, La maison des livres, 1958
(2 volumes)

BOUIKEN BAHI (Amar Abdelkader)
Dictionary of proverbs and proverbial sayings
Oran , LAROS,2007, 245p+198 p

CHEBEL (Malek)
Dictionaries of Muslim symbols. Rites, mysticism and civilisation
Paris, Michel Albin, 2000,501p

CHEVALIER (Jean) , GHEERBRANT (Alain)
Dictionaries of symbols. Myths, dreams, customs, gestures, shapes, figures,
colours, numbers
(Revised and expanded edition)
Paris, Robert Lafont/Jupiter ,1992,1060 p

COLLECTIVE
Encyclopaedia of Islam Leiden/Paris

COLLECTIVE (under the direction of Mohammed Ali Amir -MOEZZI)
Dictionary of the Koran
Paris, Robert Laffont,2008,981 p

COLLECTIVE (under the direction of Pierre BONTE and Michel IZARD)
Dictionary of ethnology and anthropology
Paris, PUF,2007,842 p

COLLECTIVE (Under the direction of A.L. de PREMARE)
Langue et culture marocaines. Dictionnaire Arabe- Français Paris, L'Harmattan,
1993(12 volumes)

GAID (Tahar)
Dictionnaire élémentaire de l'Islam Alger, OPU,1991,418p
(2ᵉ edition)

MAHREZ(Amine)
Reasoned glossary of French words of Arabic origin
Alger,éditions Dar Othmania,2006,187 p

PONT-HUMBERT (Cathérine)
Dictionary of symbols, rites and beliefs
Paris, Hachette, 2003,434 p

WARING (Philippe)
Dictionary of omens and superstitions
Translated and adapted by Christel Rollinat Paris, Editions du Rocher,1982,271p

BIBLIOGRAPHICAL WORKS

COLLECTIVE

Mille et un livres sur le monde arabe Paris, Maison des Sciences de
l'Homme,1984,290 p

COLLECTIVE

The Arab and Muslim world in the mirror of French universities. Directory of
theses defended in French universities, in Human and Social Sciences, on the
Arab and Muslim world (1973-1987)
Jacqueline Quilès, with the collaboration of Danièle Bruchet, Marie Burgat,
Monique Codron and Jean-pierre Dahdah)
Aix, CNRS, 1991,198 p

HENRY(Jean-Robert)and BALIQUE (François)

The colonial doctrine of Algerian Muslim law. Systematic bibliography and
critical introduction.
(CNRS and Centre Régional de Publications de Marseille, Les Cahiers du
CRESM) Paris, CNRS,1979,178 p

JANIER (Emile)

1- "Bibliography of the works o f   Alfred Bel
in Revue Africaine, 1945, pp110-16
2- "Bibliography of publications on Tlemcen and its region".
in Revue Africaine, 1949, pp314-34

LARNAUDE (Marcel)

"Bibliographie Algérienne (1934)" in Revue Africaine, 1935, pp196-209

MAYNADIES (Michel)

Bibliographie Algérienne. Répertoire des sources documentaires relatives à
l'Algérie Algiers, OPU, 1989,336 p

SHINAR (Pesach)
Contemporary Maghreb Islam. Annotated bibliography
Paris, Editions du CNRS, 1983, 506 p

WORKS WORKS

ABOU BEKR (Abdeslam)
1- "Usage de droit coutumier dans la région de Tlemcen", Revue Africaine,
1936, pp813-66
2- "Notes sur les amulettes chez les indigènes algériens", Revue Africaine, 1937,
pp309-18

ADDA BACHIR (Abdelkader Mazari)
The pilgrimage
Oran, Editions Dar El Adib,2006,166 p

ADHAHABIY (The Imam)
The deadly sins. Summary of the work of Imam Adhahabiy
Translation by Youssef Sattay. Revised and corrected by the translation
department Brussels, Ed. El- Fajr, 2006, 94 p

AL -JILI (Abd Al Karîm
Universal man
Translated extract from the book Al-Insan Al-Kâmil Translation from Arabic
with commentary by Titus Burckhardt
Paris, Dervy -Livre, 1975, 98 p

AL MUQADDASI
Description of the Muslim West in the 4th= 10th century.
Algiers, Editions Carbonel, 1950, 122 p (Arabic text and French translation with
introduction, notes and four indexes by Charles Pellat).

AL YA'QUBI
Description du Maghreb en 276/889 (Extrait du " Kitab al-Buldân " Alger,
Institut d'Etudes Orientales,1962,XXVII p + 61 p
(Arabic text published after the Leiden edition -1892- by Henri Peres, Professor
at the Faculty of Letters in Algiers. Foreword and annotated translation by

Gaston Wiet, member of the Institut; preface by Georges Marçais, member of
the Institut).

ANONYMOUS
1- Types of monotheism Translation and publication of
Daroussalam; revision of Mohammed Al -Amin Ben Ibrahim
Riadh, Daroussalam,2001,32 p 2-Les piliers de la foi Translation and
publication by
Daroussalam; revision of Mohammed Al -Amin Ben Ibrahim
Riadh, Daroussalam, 2005,32 p (2nd edition)

BASSET (René)
"Mille et un contes, récits et légendes arabes by H. Massé (CR of volumes I, II
and III)
Revue Africaine, 1925, pp369-71 and 1927, pp 311-2

BEAUD (Michel)
The art of the thesis
How to prepare and write a doctoral thesis, magisterial dissertation or
undergraduate dissertation
Algiers, Casbah éditions,1999, 172 p

BEAUD (Stéphane, Florence Weber)
Field survey guide (New edition)
Paris, La Découverte, 2003,357 p

La population musulmane de Tlemcen Paris, Librairie Paul Geuthner,1908,57 p
(Extracts from the R.E.E.S.1908)

BELHALFAOUI(Mohammed)
1- Popular Arabic poetry from the Maghreb
Paris, François Maspéro,1982,206 p
2- "Le Melhoun : une production classique et une relève ",Thurath (les cahiers
du CRASC) n° 15, 2006 : Le melhoun : textes et documents (sous la direction
de Ahmed Amine Dellaï)Pp 67-88

BENABADJI (Foudil)
Tlemcen in history through tales and legends
Preface by Mohammed Dib Paris, Publisud, 2003,448 p

BENABDELKRIM AL DJAZAIRI (Mohamed, Dr)
The Islamic ruling on the marriage of a Muslim man to a non-Muslim woman
and vice versa
Sl,Sd,Se,68 p
(Translation of the Arabic text)

BENACHENHOU (A)
Connaissance du Maghreb. Notions d'éthographie, d'histoire et de sociologie
Algiers, Editions Populaires de l'Armée,1971, 388 p

BENATTIA (Abderrahman)
History of a universal language: Arabic Algiers, Editions Houma, 2006,383 p

BENHADJI SERRADJ (Mohammed)
"Pages de folklore tlemceniern. Le retour du printemps", IBLA, Tunis, 1951, pp
73- 82

BENHAMOUDA (Boualem, doctor)
1- Les clés de la langue arabe Algiers, OPU,1993,413 p(2^e edition)

2- L'origine arabe de la langue Française Paris, Dialogues éditions, 1996,131p
3- The exact of certain Spanish words Studies accompanied by de citations à
apporter au dictionnaire de l'Academie Royale de Langue espagnole Algiers,
éditions Dar El Oumma,1991,126 p

BOUAMRANE (Chikh, Dr)
A look at the culture of yesterday and today
Algiers, HCI, 2005,263p

BOUCHERIT (Aziza)
Arabic spoken in Algiers
Algiers, Editions ANEP,2006,338 p

BOUDECHICHE (Smaïl)
Quranic references. Hizb "Amma" and
"Sabbih
Algiers, Dahlab, 1995,216 p

BOUDJEDRA (Rachid)
Contemporary daily life in Algeria
Paris Hachette1971, 253 p

CAILLE (Alain)
Anthropology of the gift
Paris, La Découverte, 2007,276 p

CAILLOIS (Roger)
1- Man and the sacred
Paris, Gallimard, 2006,250 p

2- Le mythe et l'homme Paris, Gallimard, 2002,189p

CAZENEUVE (Jean)
Sociologie du rite Paris PUF1971,334p

COMBESSIE (Jean-Claude)
Method in sociology
Algiers, Casbah éditions,1998,123 p

COPANS (Jean)
The survey and its methods. Ethnological fieldwork
Paris Armand Colin,2005, 137 p

CUSSET (Pierre- Yves)
The social link
Paris ,Armand Colin, 2007,126 p

DEJEUX (Jean)
Djoh'a, yesterday and today Sherbrooke, A.Naaman,1978,121 p

DERMENGHEM (Emile)

"Tlemcen mystique. Saints et confréries Tlemcen et sa région", special issue of
Richesses de France, 1er quarter 1954, pp13-20pp53-7

DJEGHLOUL (Abdelkader)
1- Eight studies on Algeria Algiers, ENAL, 1986,205 p

2- Eléments d'histoire culturelle algérienne Algiers, ENAL, 1984,244 p

EL HASSAR (Benali)
Tlemcen. City of the great masters of Arabo-Andalusian music
Alger, Editions Dalimen,2002,163 p (Preface by Mahmoud-Agha Bouayed)

ELIMAM (Abdou)
Le Maghribi
 AKA "ed-darija
(La langue consensuelle du Maghreb) Oran, Dar El Gharb,2003,237 p

**GAUDEFROY -DEMOMBYNE (M) and ZENAGUI
(Abdelaziz)**
"Récit en dialecte tlemcenien", Journal Asiatique, July-August 1904, pp45 - 117

HACHELAF (Mohamed El habib)
El Haoufi . Chants de femmes d'Algérie Algiers,éditions Alpha,2006,373p

HAMIDOU (Abdelhamid)
1- "Aperçu sur la poésie vulgaire de Tlemcen. Les deux poètes populaires de
Tlemcen : Ibn Amsaïb et Ibn Triki", Revue Africaine, 1936, pp1007-46
2- "Devinettes populaires de Tlemcen" Revue Africaine, 1937, pp 357-72

MAUSS (Marcel)
1- Ethnography manual
Paris Payot et Rivages, 2002,363 p
2- Works.1.The social functions of the sacred
Paris, Les Editions de Minuit, 2005,636p
3- Essay on the gift, followed by rapports de la psychanalyse et de la sociologie

(Reports on psychoanalysis and sociology)
Algiers, ENAG éditions, 1989,231 p (Presentation by Houria Benbarkat)
4- Essays in sociology
Paris, éditions de Minuit, 1971,252 p
5- Sociologie et anthropologie Introduction by Claude Levy Strauss. Paris, PUF, 1968, 482 p

NACIB (Youssef)
Eléments sur la tradition orale Algiers, SNED,1982,143p
(2e edition)

NEFZAOUI (Sheikh)
The Perfumed Garden. A manual of Arabic erotology.
Paris, Ina-Yas, 1999,294 p

SOUALAH (Mohammed)
1- Complementary course in spoken Arabic. Religious society: festivals, ceremonies, basics of Islam, Rites, Marabouts. The dialects of Algeria and Morocco: customs, habits, institutions of the natives, lecture plans.
Algiers, La Thypo-Litho et J.Carbonel,1958,VI p + 217 p

2- Elementary course in spoken Arabic. Reading, writing, language exercises, object lessons, stories, anecdotes, intelligence exercises.
Algiers, La Thypo-Litho et J.Carbonel,1956,VII p + 175 p

TALEB (Mohammed Nour Eddine)
"Essai d'introduction à une étude étymologique du dialecte algérien", in Annales de l'Université d'Oran, n°1, 1995, Université d'Oran, pp83-94.

VAN GENEP (Arnold)
Rites of passage.
Paris, E.Noury , 1981 reed. 226 p.

YELLES -CHAOUCH (Mourad)
Le Hawfi. Women's poetry and oral tradition in the Maghreb
Algiers, OPU, 1990,424 p

ZERDOUMI (Nafissa)
Children of yesterday. The education of children in a traditional Algerian environment
Foreword by Maxime Rodinson
Paris, François Maspero, 1982, 302 p

ABDERAHIM GUENAOU

Abderahim was born on 2 September 1990 in Ain El Hûts, where he spent his entire childhood and continues to live in the Tlemcen hawz, into a very modest family whose father, Benaissa, was a simple worker, earning his wages to provide for his small family. When he reached school age, he enrolled at the Ain El Hûts mixed school, where his father, mother, aunts and uncles, both maternal and paternal, continued their education.

His maternal grandfather was willing to continue his schooling at the school that had just opened its doors on 1 October 1938, but his parents' difficult socio-economic conditions could not be conducive to education, whether at the Koranic school or the state school, in the aftermath of the celebration of the centenary of the French occupation of Algeria and before the outbreak of the Second World War. These conditions were similar for his grandparents, on both his mother's and father's sides.

Young Abderahim continued his education until he passed the entrance exam for sixth form (2000), when he enrolled at the Collège d'Enseignement Moyen: CEM El Habbek in Abou Tachfine for a single term before moving on to the CEM Malek Bnou Anès in Négrier, now Chetouane. He passed into the first year of secondary school with the Brevet d'Enseignement Moyen, a bilingual diploma obtained in 2006.
He then went to the Lycée des trois frères Attar in Negrier (Chetouane). He studied Mathematical Technical Sciences, a speciality commonly known as Mathéleme, for one year.

In fact, his maternal grandfather insisted that he should go to secondary school so that he could follow in the footsteps of his maternal uncles. He obtained his baccalauréat T.M. (Techniques Mathématiques) in 2011.

He then continued his studies in geology, graduating in 2014 with a Bachelor's degree, followed by a Master's degree (2019). In the Hydrology speciality and defended in October 2020, the Master's thesis is entitled :

"Calculation and validation of a rainfall-discharge model: application to data from ued Isser (Tafna, north-west Algeria).

APPENDICES

Annex n° 01

The legend of water in Tlemcen A whole mythology is woven around water and snow in Tlemcen, whose name comes from "thala m'sen", which in Tamazigt means "fountain with two springs", or "thala msan", which means "gentle fountain".The legend of water in Tlemcen, told by writer Foudil Benabadji, vice-president of the "Friends of Tlemcen", is a good illustration of this hydric etymology.The country (bled) is in dire straits: El Ourit, El Mefrouch, Mouillah Boukiou, saquiet el nosrani, sahridj bedda... have all run dry. "The water has to come back to Tlemcen. I'll go and get it.This was the challenge taken up by the Prince who defeated "El Ghoul", the evil ogre. Water began to gush from the fountains like crystal pearls... The springs came back to life, like Aïn el Ouzir, Aïn bent Soltane, Aïn Keubba, Aïn Tolba, Aïn el Hout, Aïn el Hdjel, Aïn el Modj-arra (El Mouhadjir), Aïn Djnan, Aïn Sidi Ahmed, Aïn Attar... The "haoufi" songs resounded once again... in the town. And it was from that time on that our good town took the name of "Tilimcène", which means "sources" in Berber, says the storyteller...Far from the legend, the people of Tlemcen used to organise rogations to ask for rain ("tol el latif") when the lack of rain jeopardised the harvest or grazing in spring; they would especially visit the city's main patron saints, Sidi Daoudi, Sidi Boumediene, Sidi Abdelkader, Lalla Setti, Sidi Boudjemaâ... and make invocations to them. At the same time, they carried out the following tasks at the "mçalla" in Mansoura, Ouzidane, "çalat el istisqa'" (collective prayer in the open air to invoke the rain). Meanwhile, the children were not idle. They roamed the streets carrying a rag doll or scarecrow (two crossed reeds covered with an old worn-out dress) called "Ghanja" and singing Ghanja, Ghanja, natalbouk r'ja, ya rabbi aâtina ch'ta..." (Ghanja, Ghanja, fulfil hope!) (Ghanja, Ghanja, fill the hope! O my God, give us rain! And you, bells, ring out so that the poor woman without a husband may live). They would knock on the doors of the houses to be sprinkled with water as a good omen. The one with the white complexion is the meaning of "Ghanja", which would be the Berberisation of the word Ganus or Janus, a pagan etymology. A semantic shift "In Arabic, ghanja meant song, or sung poetry, in reference to the Arabic root "ghnâ" (song)...We would also sing "Ya ch'ta sabi sabi, m'a t'sabich aliya, hata dji hamou khouya, ghatini ba' zarbiya"... (Oh rain, fall, fall but don't drench me, until my brother Hamou comes to cover me with a carpet). O'urs dib" (the marriage of the fox) was celebrated "under" the rainbow when there was a combination of rain and sunshine:"chrika djat Allah, Allah, f'dila djat, Allah, Allah!" (the concubine arrives, the favourite arrives), an allusion to the rain/sun pairing. When the weather was bad and to ward off bad luck, the old women would recite a mystical lament while carding or "shelling" the "m'qatfa": "Ghitna, ghitna ya latif bi khal'qihi, ida nazala el qada' yaatba'ou loutfou" (O my God, bring Your help, You who are Temperate towards Your creatures, mitigate fatality with Your temperance).In the old days, when the cycle of the seasons was regular and the climate was free of any greenhouse effect, we lived through winter, in this case with its own intrinsic weather and cultural rituals. Source unknown

The source that deserves a mention

TABLE OF CONTENTS

BY WAY OF PRESENTATION 2

WATER (EL MA/ EL MA-E)[2] 4

THE PUBLIC FOUNTAINS OF AIN EL HUTS[3] 9

THE SPRINGS AROUND AIN EL HUTS 15

RECOMMENDATIONS FOR STAYING

HYDRATED[8] 24

PROHIBITIONS SOCIO-CULTURAL 27

BIBLIOGRAPHY 28

BIBLIOGRAPHICAL WORKS 30

APPENDICES 40

I want morebooks!

Buy your books fast and straightforward online - at one of world's fastest growing online book stores! Environmentally sound due to Print-on-Demand technologies.

Buy your books online at
www.morebooks.shop

Kaufen Sie Ihre Bücher schnell und unkompliziert online – auf einer der am schnellsten wachsenden Buchhandelsplattformen weltweit! Dank Print-On-Demand umwelt- und ressourcenschonend produziert.

Bücher schneller online kaufen
www.morebooks.shop

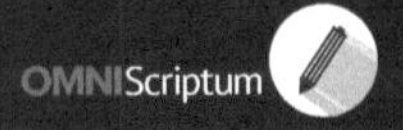